EXPLORER SERIES

WORLD'S Cutest

Baby Animals

Susanne Blumer

ISBN 978-0-9966164-6-1

Printed in the United States of America
First Printing, 2016

Want to know when a new book in the Explorer series comes out? Become an Insider at www.susanneblumer.com. You will be the first to know about upcoming books, events and free goodies!

Chickadilly Press

Our world is full of amazing animals. Are you ready to see some of them and learn a few fun & interesting things too?

Awesome! Let's go!

Giraffe

Giraffes are one of the ***tallest*** animals in the world. They can be up to 18' tall! Just their leg is taller than most people.

A baby giraffe is called a ***calf***.

A girl giraffe is called a ***cow***.
A boy giraffe is called a ***bull***.

Giraffes live in the ***savannas*** and woodlands of Africa. A savanna is a big grassy meadow.

Giraffes are ***browsers***. Browsers love to eat leaves, flowers and fruit off of trees.

Did you know?

Giraffes spend almost all of their time standing up. They even sleep standing up!

Kangaroo

Kangaroos are ***marsupials*** and are the only large animal that hops to get around!

A baby kangaroo is called a ***joey***.

A girl kangaroo is called a ***flyer*** or ***jill***.
A boy kangaroo is called a ***boomer*** or ***jack***.

Kangaroos only live in Australia.

Kangaroos are ***herbivores.*** Herbivores are animals that only eat plants.

Did you know?

Mama kangaroos have a ***pouch*** that their baby lives in for nine months after it is born. Cozy!

Penguin

Penguins are birds but they cannot fly. They love to swim and spend most of their time in the ocean.

A baby penguin is called a ***chick***.

There are not any specific names for girl and boy penguins. It is very hard to tell the boy penguins from the girl penguins.

Penguins live on every continent in the ***Southern Hemisphere***.

Penguins are ***carnivores*** which means they only eat meat. Fish is a penguin's favorite.

Did you know?

Although all penguins are black and white, there are 17 different kinds of penguins in all different sizes.

Hedgehog

Hedgehogs have pointy ***spines*** but the hedgehog's spines are not poisonous like porcupine quills.

A baby hedgehog is called a ***hoglet***.

A girl hedgehog is called a ***sow***.
A boy hedgehog is called a ***boar***.

Hedgehogs live in Europe, Asia and Africa.

Hedgehogs eat mostly insects but will eat meat and plants.

Did you know?

Hedgehogs have about 5,000 spines and each one lasts about a year before it falls out and a new one grows!

Polar Bear

Polar bears are the largest bears but they do not ***hibernate*** like other bears.

A baby polar bear is called a ***cub***.

A girl polar bear is called a ***sow***.
A boy polar bear is called a ***boar***.

Polar bears are only found near the ***Arctic***.

Polar bears are ***carnivores***. They mostly eat meat but will eat eggs and plants if no meat is available.

Did you know?

Polar bears look white but their hair is actually clear. Light reflects off the hair strands and makes them look white!

Lion

Lions are one of the biggest cats but they are very lazy. Lions sleep up to 20 hours a day! The girls do most of the hunting.

A baby lion is called a ***cub***.

A girl lion is called a ***lioness***.
A boy lion is called a ***lion***.

Almost all lions live in Africa but a few do still live in Asia.

Lions are ***carnivores***. The "King of the Jungle" eats many kinds of animals.

Did you know?

You can hear a lion's roar from as far as 5 miles away and lions can run up to 50 mph!

Panda Bear

Pandas are white with black fur on their eyes, ears, legs, arms and shoulders.

A baby panda is called a ***cub***.

A girl panda is called a ***sow***.
A boy panda is called a ***boar***.

Pandas live in the forests of China.

Pandas eat the leaves, stems and shoots of ***bamboo***. They can eat over 20 pounds of bamboo a day!

Did you know?

Pandas are ***endangered*** which means they are disappearing from the earth. There are only 2,000 pandas left in the wild.

Koala

Koalas have a pouch just like the kangaroo. Koalas are not bears though even people call them that! They are a ***marsupial***.

A baby koala is called a ***joey***.

A girl koala is called a ***doe***.
A boy koala is called a ***buck***.

Koalas live in the woodlands of Australia.

Koalas eat ***eucalyptus*** leaves. Most of their drinking water comes from these leaves.

Did you know?

Koalas sleep 18-22 hours a day and will store snacks in their cheeks!

Elephant

Elephants are one of the biggest animals in the world. A male elephant can be 13' tall and weigh 15,000 pounds!

A baby elephant is called a ***calf***.

A girl elephant is called a ***cow***.
A boy elephant is called a ***bull***.

Elephants live in Africa and Asia.

Elephants are ***herbivores***. They can eat 350 pounds of plants, leaves and roots a day.

Did you know?

The elephant's ***trunk*** is both his nose and his upper lip! Elephants can live to be 50-70 years old.

Fox

The fox is a member of the dog family. There are over 30 different kinds of foxes but the red fox is the most common.

A baby fox is called a ***kit*** or ***pup***.

A girl fox is called a ***vixen***.
A boy fox is called a ***dog***.

Foxes live everywhere in the world except Antarctica. They even live in cities and neighborhoods.

Foxes are ***omnivores***. An omnivore eats both meat and plants.

Did you know?

Foxes can make 40 different sounds including a scream, growl and yelp.

Sheep

Sheep are often raised as livestock on farms and provide meat, milk and wool.

A baby sheep is called a ***lamb***.

A girl sheep is called an ***ewe***.
A boy sheep is called a ***ram***.

Sheep can be found all over the world.

Sheep like to ***graze*** on grass in the spring and summer and will eat hay and grain in the wintertime.

Did you know?

Most breeds of sheep are ***sheared*** once a year for their wool. This wool becomes sweaters, mittens, hats and blankets!

Tiger

Tigers are the largest member of the cat family. Tigers are known for their dark stripes and reddish fur.

A baby tiger is called a ***cub***.

A girl tiger is called a ***tigress***.
A boy tiger is called a ***tiger***.

Tigers live in Southeast Asia, Russia, China and Korea.

Tigers are meat-eaters and can eat up to 88 pounds of food at one time. Wow!

Did you know?

Tigers can purr and are good swimmers.

Bear

Bears are large mammals that ***hibernate*** in a den during the winter. They are very smart with an excellent sense of smell.

A baby bear is called a ***cub***.

A girl bear is called a ***sow***.
A boy bear is called a ***boar***.

Bears live in Europe, Asia and North America.

Bears are omnivores and eat both meat and plants.

Did you know?

Bears have two layers of fur: a short layer to keep them warm and a long layer to keep water away from their skin.

Cheetah

Cheetahs are the world's fastest land animal. They can go from standing still to 60 mph in three seconds.

A baby cheetah is called a ***cub***.

A girl cheetah is called a ***cheetah***.
A boy cheetah is called a ***cheetah***.

Cheetahs live in Africa and Iran.

Cheetahs are carnivores and only eat meat. They mainly eat gazelles.

Did you know?

Cheetahs are the only big cat that cannot roar. They do purr though. Cheetahs only need to drink every 3-4 days.

Orangutan

The word ***orangutan*** means "person of the forest". They are very smart and can reason and think.

A baby orangutan is called a ***baby***.

A girl orangutan is called an ***orangutan***.
A boy orangutan is called an ***orangutan***.
Tricky, huh?

Orangutans live in the rain forests of Asia.

Orangutans are ***omnivores***. Their diet is mostly fruit and leaves but they do sometimes eat meat.

Did you know?

Orangutans love to travel by swinging from one branch to another. They sleep up in trees in a nest they make each night.

Hippopotamus

Hippos are large, round, water-loving animals. They spend up to 16 hours a day in the water.

A baby hippo is called a ***calf***.

A girl hippo is called a ***cow***.
A boy hippo is called a ***bull***.

Hippopotamuses live in Africa.

Hippos are herbivores. They eat plants and come out of the water at night to eat.

Did you know?

Hippos need to come out of the water every 3-5 minutes to breathe. If a hippo is sleeping underwater, he will rise and breathe without waking up.

Donkey

Donkeys look like horses but they are actually more closely related to the zebra. Donkeys have very big ears and amazing hearing. Their ears can almost turn all the way around like a big radar!

A baby donkey is called a ***colt*** or ***foal***.

A girl donkey is called a ***jenny***.
A boy donkey is called a ***jack***.

Donkeys can be found in most of the world including Europe, Asia, Africa, the Americas and Australia.

Donkeys are herbivores. They eat grass and plants during warm weather and hay and grain in the winter.

Did you know?

Donkeys do not run when scared. Donkeys will kick with both front and back feet and will bite. Watch out!

Glossary

Arctic The arctic is a polar region at the northernmost part of the Earth.

Bamboo Bamboo is the fastest growing plant in the world.

Carnivore An animal that is a meat-eater.

Endangered When an animal is endangered, it is at risk of extinction or not existing any longer on the earth.

Eucalyptus A plant found mostly in Australia.

Graze To eat, nibble or feed on grass.

Herbivore An animal that is a plant-eater.

Hibernate When an animal spends the winter sleeping or resting.

Marsupial A type of mammal that usually has a pouch to carry their young.

Omnivore An animal that is both a meat-eater and a plant-eater.

Savanna A rolling grassland with a few shrubs and trees. Also known as a tropical grassland.

Shear The annual removal of wool from a sheep.

Southern Hemisphere The half sphere of the Earth that is south of the equator.

Spine Hardened hairs that have a solid center.

Made in United States
North Haven, CT
14 September 2026

18576392R00024